Feng Xiao

Structural Health Monitoring

AF153312

Feng Xiao

Structural Health Monitoring

Alaskan Bridge Research in Cold Remote Area

LAP LAMBERT Academic Publishing

Imprint

Any brand names and product names mentioned in this book are subject to trademark, brand or patent protection and are trademarks or registered trademarks of their respective holders. The use of brand names, product names, common names, trade names, product descriptions etc. even without a particular marking in this work is in no way to be construed to mean that such names may be regarded as unrestricted in respect of trademark and brand protection legislation and could thus be used by anyone.

Cover image: www.ingimage.com

Publisher:
LAP LAMBERT Academic Publishing
is a trademark of
Dodo Books Indian Ocean Ltd. and OmniScriptum S.R.L publishing group

120 High Road, East Finchley, London, N2 9ED, United Kingdom
Str. Armeneasca 28/1, office 1, Chisinau MD-2012, Republic of Moldova, Europe
Managing Directors: Ieva Konstantinova, Victoria Ursu
info@omniscriptum.com

Printed at: see last page
ISBN: 978-3-659-33883-0

Zugl. / Approved by: Fairbanks, University of Alaska Fairbanks, Diss., 2012

To Ling Zhao and Anqi Liu

Acknowledgments

I would like to thank Professor J. Leroy Hulsey, for his endless support and guidance of my research throughout my study. He is my ultimate role model as a researcher, advisor and lecturer. His dedication to teaching is sincerely appreciated.

I would also like to thank Dr. Gang S Chen, Dr. Juanyu Liu and Dr. Andrew T. Metzger. Their advices were invaluable in finalizing the research effort. Their support is greatly appreciated.

My research and work on the Klehini River Bridge monitoring project would not have been possible without the support of Dr. Yongtao Dong. His endless help and input are truly appreciated. I also thank Professor He Liu for supporting me on the Klehini River Bridge project. Professor Chengfu Chen helped me on the finite element analysis.

The study was sponsored and supported by Alaska University Transportation Center (AUTC) Grant Number 510015 and the Alaska Department of Transportation & Public Facilities. Their financial supports are greatly appreciated.

I thank my parents for their encouragement for continuing my study and support during all of the challenges I have faced.

Abstract

The objective of the research is to improve the safety of bridge structures in the state of Alaska through implementation of innovative structural health monitoring (SHM) technologies. The idea is to evaluate structural integrity and serviceability, and to provide reliable information for changing structural response, etc. of monitored bridges.

Based on the finite element model's moving load analysis, modal analysis results and field inspection, this study was used to establish a bridge SHM system for a particular bridge including a preferred sensor layout, system integrator and instrumentation suitable for Alaska's remote locations with harsh weather.

A variety of sensors were proposed to measure and monitor structural and environmental conditions to assist in the evaluation of the performance of the Klehini River Bridge. This system is able to provide more reliable information on the real structural health condition. It can be used to improve safe performance of this bridge. As a new safety and management tool, this SHM system will complement traditional bridge inspection methods. Implementation of an effective monitoring system will likely result in a reduction in inspection manpower, early detection of deterioration/damage, development of optimum inspection cycle and repair schedules before deterioration/damage grows to a condition where major repairs are required.

Table of Contents

List of Figures

List of Tables

Chapter 1 Introduction

1.1 Problem Statement

There are nearly 600,000 bridges in the United States and 26% were identified as structurally deficient or "functionally obsolete" in a report from U.S. Department of Transportation. Structurally deficient or functionally obsolete bridges can pose a risk to public safety and are economically challenging. While not necessarily unsafe, a structurally deficient bridge may either be closed or restricted to light vehicles. This is typically because there are deteriorated structural components that lower the bridge rating. The same U.S. Department of Transportation report classified about 27% of the 1200 bridges (including federal bridges, otherwise 23% of 959 non-federal bridges) in Alaska as being structurally deficient and functionally obsolete. Due to the harsh environment of high latitude and heavy truck loads, the capability to monitor the condition of bridge structures on Alaska highways is important to the Alaska Department of Transportation & Public Facilities (ADOT&PF). It is particularly important that the monitoring system must have the capability to detect early stages of damage or changes in condition. Current damage detection methods rely on a biennial visual inspection by bridge inspectors, and occasional enhanced inspection using non-destructive testing/evaluation (NDT/NDE). These methods require the location, or possible location, of damage to be known prior to the assessment. Many conventional damage detection techniques are not real time and do not allow for systematic comparison of the assessment results. Consequently, damage or deterioration cannot be easily monitored or tracked over time. Real-time structural monitoring is needed to accurately evaluate and effectively manage selected bridges in Alaska.

In response to this need, structural health monitoring (SHM) has become a much discussed topic, but it has not been widely implemented yet. Bridge SHM is a process of evaluating the condition or condition change by collection and interpretation of data from sensor-instrumented bridges. In the recent past, there have been rapid advances in the technologies for bridge evaluation. If properly implemented, structural health monitoring can aid in several aspects of bridge management, such as reducing inspection costs while improving quality, prioritizing repair/maintenance schedules, and increasing accuracy of both deterioration estimations and the decision-making process. However, challenges exist in integration and interpretation of the information from sensor networks. Additional difficulties arise for the SHM of bridges in remote, cold regions such as in Alaska. This is because of the effect of the harsh environment on the reliability and durability of the SHM equipment and sensors, power supplies and data communication tools.

To address these challenges, the overall objectives for this study are to establish a SHM system based on available knowledge and technologies for bridges in cold, harsh environments. The SHM shall provide guidelines for implementation of the SHM program. Using this system, we proposed to instrument the Klehini River Bridge to monitor its structural response to active traffic loading and to evaluate its structural condition in real time. Development and implementation of a real-time SHM program for the Klehini River Bridge should greatly enhance the ADOT&PF Bridge Section's ability to safely manage this bridge during its service life.

1.2 Bridge Description

Many of America's bridges are aging and experiencing deterioration. It's very difficult to properly evaluate structural condition and performance of these bridges under today's traffic. Further as a part of this evaluation process, it is often difficult to decide which structural components need to be retrofitted or replaced by new structural members in order to optimize the available budget. As more and more state DOTs realize the importance of monitoring bridge performance, these states are developing appropriate SHM programs for their bridges. The ADOT&PF also has shown interest in the possibility of applying structural health monitoring and damage detection technologies to Alaska's bridges.

Bridges in Alaska are routinely subjected to harsh weather conditions and usually located in remote areas. Maintenance, rehabilitation and replacement of these bridges in a cost effective manner depend critically on reliable inspection and condition assessment. Inspections of these bridges are both costly and time consuming. Compared with other states in the nation, bridge monitoring in Alaska is more needed but also more challenging. This is partially due to the harsh weather conditions and issues related to remoteness. For example, power is not always available at the bridge site and thereby this causes special challenges in data retrieval and reliable data communication from remote sites.

The Klehini River Bridge is located on the Porcupine Crossing Road accessed at mile point 26.3 of Haines Highway. The bridge structure is made of two-span riveted steel parker truss (see Fig. 1.1). The total length of this bridge is 74 meters (243 feet). The dimensions of the Klehini River Bridge is shown in Fig. 1.2. The superstructure consists of various box sections with inverted channel sections riveted to two steel plates. The timber deck is supported by a series of timber girders connected to transverse I-beams. Both spans rest on a central concrete abutment and the side banks.

Figure 1.1 Klehini River Bridge (Photo Courtesy of ADOT&PF)

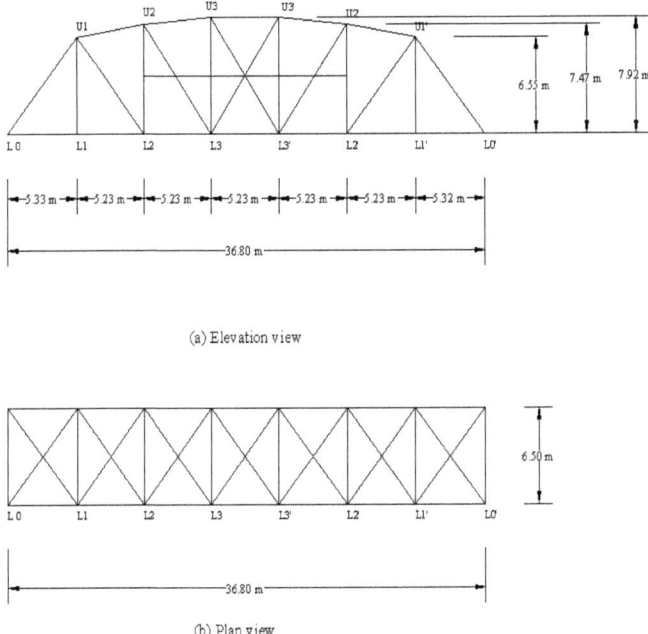

(a) Elevation view

(b) Plan view

Figure 1.2 Bridge Dimension

The boundary conditions for the girders are shown in Figure 1.3. There are four fixed supports in the center and four expansion supports in the north and south sides of the bridge.

(a) Expansion Support

(b) Fixed Support

Figure 1.3 Boundary Conditions

The truss structure originally spanned over the Mendenhall River in Juneau and was known as the

Mendenhall River Bridge. In 1969 and 1971, the trusses were partially disassembled, shipped to

Haines, and installed at their current location. As defined by the Federal Highway Administration, the Klehini River Bridge is structurally deficient with the following National Bridge Inventory condition ratings:

- deck – 7 (good)
- superstructure – 3 (serious)
- substructure – 4 (poor)

The poor superstructure condition rating is based on observed damages to the steel trusses, many of which are believed to have been caused by demolition, shipping, assembly or a combination of those activities during the 1969 and 1971. The truss damage has been noted since a 1974 inspection. Recent ADOT&PF inspections [1, 2] reported the current damage in a variety of structural members which include torn gusset plates, cracking at rivet holes, damaged or missing lateral bracing, damaged sway bracing, and etc. Examples of these conditions are shown in Fig. 1.4 to Fig. 1.8. Weld repairs were also identified at several locations of the structural elements. Gouges, flame cut holes, bullet holes, and tack welds for cracks on the truss members were also great concerns to potential degradation. Although the Statewide Transportation Improvement Program (STIP) indicated that construction funds to replace the bridge will be available in 2013, it is not unreasonable to expect that the bridge will remain in-service through 2015. Thus, it needs to be monitored to assess safety and performance until it is replaced.

| Bridge No. | 1216 | Br Name | Klehini River | Date | 07/08/08 |
| Roll No. | 1 | Inspector | Sielbach/Orbistondo | Frame | 66 |

Span 1, L1' Tear In Upstream NE Gusset Plate

Figure 1.4 Torn Gusset Plate (Photo Courtesy of ADOT&PF)

| Bridge No. | 1216 | Br. Name | Klehini River | Date | 07/08/08 |
| Roll No. | 1 | Inspector | Sielbach/Orbistondo | Frame | 79 |

Span 2, upstream truss: Crack in member L0'-L1' at L1'

Figure 1.5 Cracking at a Rivet Hole (Photo Courtesy of ADOT&PF)

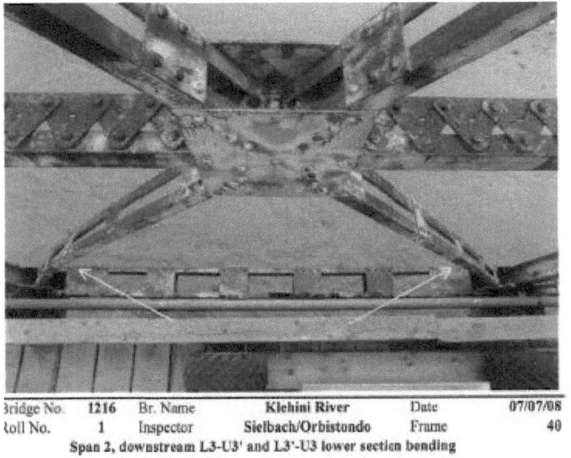

| Bridge No. | 1216 | Br. Name | Klehini River | Date | 07/07/08 |
| Roll No. | 1 | Inspector | Sielbach/Orbistondo | Frame | 40 |

Span 2, downstream L3-U3' and L3'-U3 lower section bending

Figure 1.6 Bent Truss Members (Photo Courtesy of ADOT&PF)

| Bridge No. | 1216 | Br. Name | Klehini River | Date | 05/05/10 |
| Roll No. | 3 | Inspector | Higgs/Levings | Frame | 31 |

Bent Sway Brace at U2 span 1 (From DS Side)

Figure 1.7 Bent Lower Sway Bracing (Photo Courtesy of ADOT&PF)

| Bridge No. | 1216 | Br. Name | Klehini River | Date | 07/08/08 |
| Roll No. | 1 | Inspector | Sielbach/Orbistondo | Frame | 73 |

Span 2, lower lateral bracing US L1 to DS L0 Heavily Bowed

Figure 1.8 Damaged Lower Lateral Bracing (Photo Courtesy of ADOT&PF)

Since it serves as the only access route to this region, the Klehini River Bridge is a vital link to several small communities in the area. ADOT&PF is tasked with inspecting and managing the bridge until it is replaced. To accomplish this, the Bridge Section at ADOT&PF performs annual hands-on inspections supplemented with NDE using magnetic particle and ultrasonic examination [1, 2] on previously identified deformations, defects or welded repairs. These inspections only provided temporary condition evaluations at known defective areas. To ensure safe operation of this bridge, the footprint of the bridge response to active traffic and the capability of detecting changes in current defective areas on a more frequent basis are needed.

1.3 Structural Health Monitoring Technology

Recent advances in structural health monitoring technologies have made it possible to monitor the structural response when subjected to active loadings or changes to the structure over time. Therefore these tools can be used to determine the necessary maintenance/repair, retrofit or strengthening, and to predict possible failure [3, 4]. However, current implementation of SHM on bridges is not common because there is no standard procedure to follow. Challenges also exist in integration and interpretation of the information from SHM systems. Additional difficulties arise for SHM of bridges in remote, cold regions of Alaska. Such issues as the effects of the harsh environment on the sensors and equipment, the power needs at a remote site and the demand for data communication from a remote bridge site should be investigated before providing recommendations for a SHM program for bridges in these types of regions.

There are a variety of sensors available today ranging from fiber optic systems to dielectric and piezoelectric sensors, to strain gauges, MEMS (micro-electro-mechanical systems) sensors, MOTES (wireless sensors with capabilities of performing some processing, gathering sensory information and communicating with other connected nodes in the network) and even lasers. The choice of the sensors will depend on the needs of measurement (for example, the stress, deformation/movement, cracks, corrosion, etc.), the required sensitivity, the operational environment and the long-term durability and reliability. This is of special importance when sensors are bonded or otherwise placed on surfaces, which may be subject to large temperature variations (both daily and over seasons), moisture (ranging from humidity and precipitation caused by immersion such as when flash floods cause overtopping of bridges), and UV radiation.

Wireless connections are developed to meet needs for some large structures where lead cable transmitted sensor signals might be corrupted by excessive noise, or where long lead cables are otherwise impractical. These systems are also of value in areas where vandalism may likely cause damage to a wired system. Wireless sensor networks can be used to acquire data from sensors to on-site DAQ instruments. Wireless data transfer is currently more expensive than direct connections. That is, the data is typically transferred much more slowly, and the signals are not completely secure. However, in the future, it is expected that wireless communications will become more popular and more reliable for SHM on very large structures. Using wireless technology is especially beneficial when the monitored points are not easy to access and running wires/cables are difficult to connect the sensors to data acquisition instruments.

While a number of new sensor systems have demonstrated good short-term durability with the ability to ensure a high level of self-monitoring and self-calibration, their long-term durability under severely changing environment with the low levels of conventional maintenance/inspection seen in civil infrastructures, is yet unknown. Further, most sensing systems drift with time; this is a problem that is yet to be solved satisfactorily.

Similar to the challenge of sensor selection and placement, is the challenge of assessing valid data. While it is useful to be able to both see the data stream in real-time and to archive them for future use, the real purpose of the data in SHM is to allow interpretation and analysis. Unfortunately, without good planning the SHM system may result in being an intricate measure of collecting data, rather than to provide a means for efficient management and interpretation. It is therefore critical that the system provides a means not just of recording (and displaying) response, but also (and more importantly) of characterizing the response and comparing it to an appropriately updated model to enable assessment of the critical aspects of capacity and service-life.

When configured with telemetry, data can be transmitted in real-time from the bridge site to a project computer, eliminating the need for periodic site visits to upload data. Two factors help determine the best telemetry method - site conditions and distance to the project computer. When the project computer can be located within a few miles (line-of-site) or a few hundred feet (non-line-of-site), license free spread spectrum radio telemetry is the best choice. If the project site is remotely located relative to the project computer, cellular or satellite telemetry becomes the best option. If the site has access to a landline or Ethernet hub, then options are also available for landline phone or Ethernet telemetry.

Despite these issues and challenges, tremendous advances have been made in the actual development and implementation of SHM systems for civil infrastructures [5~10]. Based on the ability to acquire data, transmit it, interrogate it, and then make decisions based on the cumulative sets of data stored in the database, the SHM systems are capable of serving as true tools for health monitoring – i.e. not just being able to state that the "patient" is sick, but rather being able to pinpoint location and reason, as well as the effect of the incapacity. The structural health monitoring procedure is often defined in terms of levels [11]:

Level I: Identify that damage has occurred.

Level II: Identify that damage has occurred and determine the location of damage.

Level III: Identify that damage has occurred, locate the damage, and estimate its severity.

Level IV: Identify that damage has occurred, locate the damage, and estimate its severity, and evaluate the impact of damage on the structure or estimate the remaining useful life of the structure.

Therefore the SHM system is in another meaning a decision system that is fronted by sensors and backed by knowledge base.

As part of this study, the researchers examined alternatives for providing hand held data inspection methods that are sufficient for use by the bridge inspector. Possible example of these types of systems include being able to interrogate the condition of prestressed strands or evaluate locations of critical stress, bearing loads or perhaps deck condition and states of corrosion [7].

The established SHM system was implemented to the structurally deficient Klehini River Bridge. The monitoring system can provide better tools to evaluate performance of the entire structure and the damaged elements, response of the defects during usage, and appearance of new degradations.

Chapter 2 Research Approaches

2.1 Structural Characterization

Structural characterization was conducted for the purpose of obtaining information of on status of the Klehini River Bridge just prior to sensor installation. This includes recent inspections reports and NDT/NDE results, details regarding any significant maintenance activities or modifications to the structure, and numerical modeling and analysis of the structure to simulate the loading effects and the structural response.

In modeling and analysis, the 3-D finite element models of the Klehini River Bridge were developed by using the structural analysis program, SAP2000 and ABAQUS. A global model was based on the original bridge condition according to the original construction document and the as-built condition since inspection has shown the variance of the as-built members from the construction drawings. Based on the moving load analysis and the modal analysis, the global model determined the sensor layout and modal characteristics. A local finite element model was developed by using ABAQUS. This model was used to figure out the stiffness of semi-rigid connections.

2.2 Measurement Needs

The information needed for monitoring Klehini River Bridge includes: (a) real-time quantitative information on the bridge's response to live load and environmental changes, (b) the peak compression strain and the peak tension strain not exceed the design limit, (c) the existing damage/defects and their propagations, (d) evaluation of the integrity and behavior of the structure, and (e) the effects due to occasional heavy truck loads that could cause possible damage and fatigue. The parameters to be measured are strain and displacements/deflections in critical or damaged members and joints, crack width and its growth, movements, vibrations and acceleration, environmental parameters (such as temperature, humidity, precipitation), etc. Some parameters are static in nature while others dynamic. Continuous long-term data on the strain and deformation, etc. are especially needed.

2.3 Types of Testing and Monitoring

Measurement parameters require both static and dynamic testing as part of the SHM program to obtain data for structural health assessment. The non-destructive field static and dynamic testing can be performed by placing a loaded and calibrated ADOT&PF dump truck at known locations on the bridge, followed by driving cross the bridge at a crawl speed to simulate static loads versus location. This would be followed by having the truck drive over a bump of known height on the bridge. The purpose is to induce vibration to the bridge structure. Vibration equipment can also be used as an alternative to excite the bridge structure in dynamic testing. From the dynamic test, natural frequencies and mode shapes of the bridge can be extracted and will be compared with the analytical results from the "current in use" structural model of the bridge. The finite element model will be calibrated by field static and non-destructive dynamic testing. The calibrated model will be more representative to the real bridge condition and can be used to predict the response of the bridge to active traffic loading.

In order to meet the identified monitoring objectives, the established SHM system can be used to monitor the bridge response either periodically or continuously. Periodic monitoring is conducted to investigate structural response for the purpose of determining change that might occur in the bridge at specified time or time intervals (e.g. weeks, months, or years apart). Analysis of the data may indicate damage or deterioration. For example, monitoring static field testing or moving traffic, monitoring crack growth, monitoring before and after a repair, can all be done periodically.

Continuous monitoring refers to monitoring of the structure for an extended period of time (weeks, months, or years) without interruption. In continuous monitoring, data acquired at the structure are either collected or stored on site (logged) for transfer, analysis, and interpretation at a later time, or they are continuously communicated to an offsite (remote) location. The established SHM system is capable of transmitting field data remotely to the engineer's office for real-time monitoring and interpretation. Since the superstructure of the Klehini River Bridge is in serious condition, it is necessary to apply continuous monitoring for assessment of its structural integrity and safety under routine traffic loadings. On the other hand, continuous monitoring is sophisticated due in part to the higher costs and difficulty of management of immense amount of data. This type of monitoring may be commissioned as especially needed when seasonal higher volume of traffic passes the bridge. As an alternative, it may be sufficient to establish thresholds or data flags that indicate the behavior may change due to the subject load condition.

2.4 Selection of Sensors and Systems

Based on review and evaluation of current SHM technologies and systems, the research team decided to use fiber optic sensors (FOSs) and a data acquisition system (DAQ). Compared to conventional strain gages, FOSs offer the following advantages [12, 13]:

- **Stability:** light signals can be transmitted along very long lengths with a very low signal transmission loss, allowing remote monitoring. FOSs are free from corrosion, having long-term stability, and allowing continuous monitoring;

- **Non-conductive:** FOSs are free from electromagnetic and radio frequency interferences, avoiding undesirable noise;

- **Convenience:** FOSs and cabling are very small and light, making it possible to permanently incorporate them into the structures. Long gage sensors are available for distributed sensing and the sensors can be virtually applied to any structural shape.

A number of recent bridge SHM projects have been using fiber optic sensing technologies and there are a quiet few fiber optic sensor manufacturers who provide not only sensors for strain, displacement, acceleration measurements, but also fiber optic interrogator and data management software [10].

Fiber-Bragg grating (FBG) optical sensors include: single or serialized FBGs in polyimide coated fiber, FBG strain and displacement sensors (regular and rugged), FBG accelerometers, and FBG temperature sensors. These sensors can be used in an environment of -40°C to 120°C (up to 80°C for accelerometers) with instrumentation in an environment-controlled enclosure. Data is transferred to a central system for further processing and analysis. All the sensors will be fiber optics based and therefore only one sensor interrogator system is required for this project.

Chapter 3 Development of the Structural Health Monitoring System

3.1 Introduction

This study addresses specific issues associated with the bridge in question, i.e. torn gusset plates, cracks at rivet holes, damaged or missing lateral bracing, damaged sway bracing, and the soundness of identified weld repairs on structural elements at several locations. The proposed monitoring plan includes extraction modal characteristic using accelerometers, and local diagnostic monitoring through the use of strain and crack gauges.

Since damages and deteriorations exist at many locations on the bridge, it is impractical to install sensors at all locations that are damaged. Therefore, optimization of the sensor layout for the select bridge was based on the results of moving load analysis, modal analysis and the latest inspection reports.

3.2 Moving Load Analysis

Three-dimensional linear elastic finite element global models of Klehini River Bridge have been constructed in SAP2000 (Fig. 3.1), a finite element analysis computer program. The model represents the structure in its current as-built configuration. The truss members, girders, stringer and floor beams were modeled by frame elements that have three translational degrees of freedom (DOFs) and three rotational DOFs at each node. The deck was modeled by shell elements.

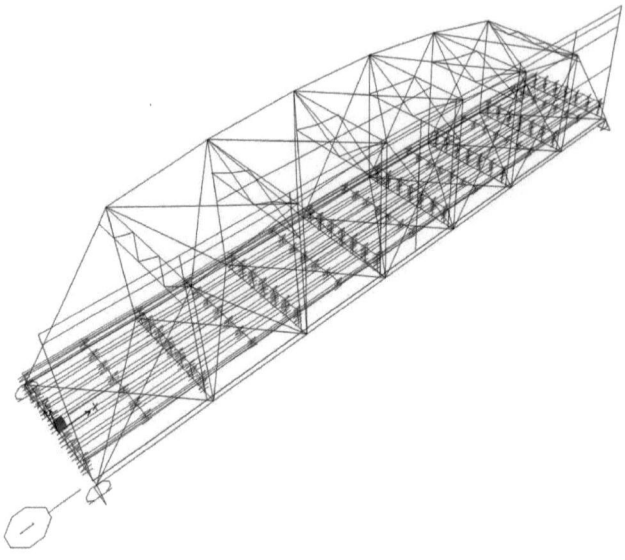

Figure 3.1 Global Finite Element Model

Most of the finite element global models for truss structure assume the connection as hinged, but the actual condition of the connection is the semi-rigid connection. In order to estimate the worst condition of connections' influence on the critical members, three finite element models (Model-1 Model-2 and Model-3) were developed. In finite element Model-1, the truss's connections were assumed as hinge connections. In finite element Model-2, the truss's connections were assumed as rigid connections. In finite element Model-3, the truss's connections were assumed as hinge connection and the poor support conditions at abutment (Fig. 3.2), including oxidation and soil build up, were considered in the model. The worst condition is when the expansion bearings are not free to rotate. That assumed the expansion bearing to be a fixed bearing.

Figure 3.2 Partly Buried Expansion Bearing

Bridge bearings were modeled using rigid elements to connect the superstructure and pier to simulate the actual behavior. The fixed bearing behavior at a pier was modeled by simply releasing the rotational DOFs in the vertical bending plane of the bridge. For Model-1 and Model-2, the expansion bearing behavior at the abutment was modeled by assigning roller restraints in the longitudinal direction and hinge restraints in the transverse direction at the bearings. In other words, the DOFs allowed are the longitudinal translation and the vertical bending rotation. For Model-3, the expansion bearing behavior at the abutment was modeled using fixed bearing behavior. This approximates the poor support conditions.

Moving Load analysis results were based on the three models discussed above. The finite element global models can figure out critical section of the bridge by using a moving-load analysis in SAP2000. There is only one traffic lane on the Klehini River Bridge. The vehicle class was defined to contain three types of vehicle HL-93K, HL-93M, and HL-93S (Figure 3.3). Vehicles moved in both directions along one lane of the bridge. The program was used to evaluate the maximum and minimum response

quantities throughout the structure due to placement of different vehicles.

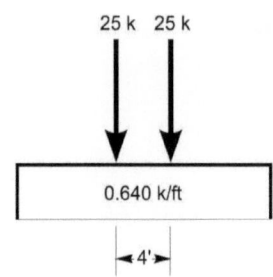

(a) HL-93M Tandem and Lane Load

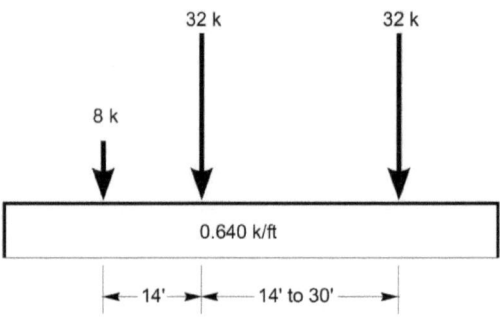

(b) HL-93K Truck and Lane Load

Figure 3.3 AASHTO Standard HL Vehicles (Images Courtesy of AASHTO)

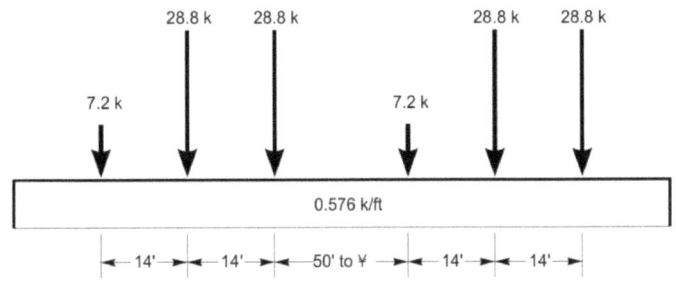

(c) HL-93S Truck and Lane Load

Figure 3.3 Continued AASHTO Standard HL Vehicles (Images Courtesy of AASHTO)

Axial strains become obvious for the north and south side truss members (Figures 3.4 and 3.5). Strain distributions are similar for the rigid connection and the pin point connection, so the rotational stiffness does not appear to influence strain distribution.

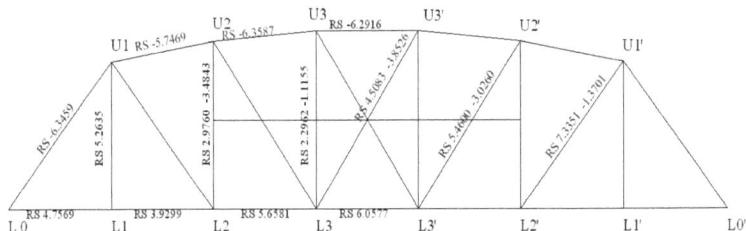

Figure 3.4 Model-1 Hinge Connection Kip/in^2 (RS: Release Section)

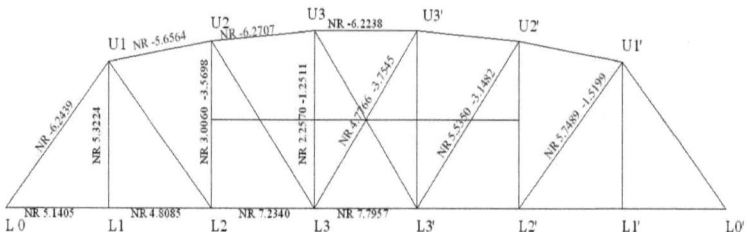

Figure 3.5 Model-2 Fixed Connection Kip/in² (NR: Non Release)

The axial strains for the poor support condition model (Fig. 3.6) show that nearly half of the lower chord load was transferred to the upper chord. However, peak strain positions did not change.

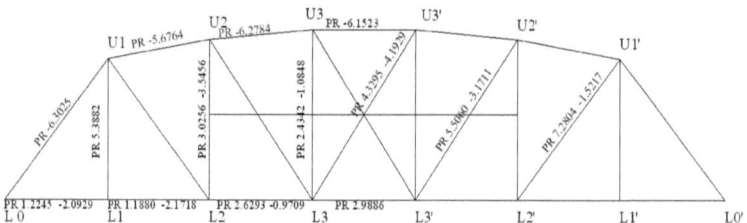

Figure 3.6 Model-3 Poor Support Condition Kip/in²

(PR: Poor Support and Release Section)

The strain gauges will be used to provide a stress history of the members. This is used to assess if the members are over stressed and if there are any bending stresses in these members. The strain diagrams for the different models provide a priority arrangement for the strain gauges (Table 3.1). Axial stresses

are classified into four cases: lower chord, top chord, diagonal and vertical.

Table 3.1 Peak Load Rating

Peak Load Rating		Mode-1	Mode-2	Mode-3
Lower chord 1	Member	L3-L3'	L3-L3'	L3-L3'
	Strain (Kip/in^2)	6.0577	7.7957	2.9886
Lower chord 2	Member	L2-L3	L2-L3	L2-L3
	Strain (Kip/in^2)	5.6581	7.234	2.6293/ -0.9709
Top chord 1	Member	U2-U3	U2-U3	L0-U1
	Strain (Kip/in^2)	-6.3587	-6.2707	-6.3025
Top chord 2	Member	L0-U1	L0-U1	U2-U3
	Strain (Kip/in^2)	-6.3459	-6.2439	-6.2784
Diagonal 1	Member	L2'-U1'	L2'-U1'	L2'-U1'
	Strain (Kip/in^2)	7.3351/-1.3701	5.7489/ -1.5199	7.2804/ -1.5217
Diagonal 2	Member	L3'-U2'	L3'-U2'	L3'-U2'
	Strain (Kip/in^2)	5.4600/ -3.0260	4.7766/-3.7545	5.060/ -3.1711
Vertical 1	Member	L1-U1	L1-U1	L1-U1
	Strain (Kip/in^2)	5.2635	5.3224	5.3882
Vertical 2	Member	L2-U2	L2-U2	L2-U2
	Strain (Kip/in^2)	2.9760/ -1.1155	3.0060/ -3.5698	3.0256/ -3.5456

Peak compression strain and peak tensile strain are essential to monitoring. Sections having both

large tension and compression also need to be monitored in that the failure stress is significantly lower

than others. Peak compression strain appears at the end of top chord. Peak tensile strain appears at the

middle of lower chord. Peak compression-tensile appears at the outside of diagonals. Preliminary strain

sensor layout is shown in Figure 3.7.

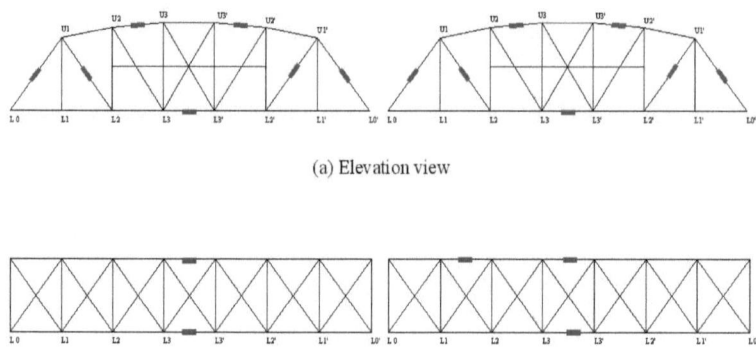

(a) Elevation view

(b) Plan view
Figure 3.7 Preliminary Strain Sensor Layout

3.2 Modal Analysis

Modal analysis can be used to determine the actual stiffness of this bridge. Stiffness matrices are dominated by higher modes and flexibility matrices are dominated by lower modes. So the actual stiffness of bridge can be identified by adjusted stiffness matrices until finite element model's higher modes equal to measured modes. In order to measure higher modes and determine the natural frequencies, an accelerometer sensor plan should be chosen based on the initial finite element modal analysis results. The positions of accelerometers depend on the lower mode shapes in longitudinal, transverse, vertical and rotational directions.

In a finite element modal analysis, natural frequencies, mode vectors and mass participation factors were determined by the Ritz–vector method. The mass participation factor for a mode provides a measure of how important the mode is for computing the response to the acceleration loads in each of the three global directions. In building design, there is a rule of thumb that the accumulated modal mass participation factor in every direction is over 90%. An analysis of the bridge specified a need for a total of 120 modes to achieve this percentage

The natural periods and mass participation factors for the first 50 modes are presented in Table 3.2 along with a brief description of dominant motion. In Table 3.2, UX, UY, and UZ represent the motions in traffic or longitudinal direction, lateral or transverse direction, and vertical direction, respectively. RX, RY and RZ stand for the rotation in each direction. Description is used to indicate mode shapes corresponding to each direction. The natural period of the bridge ranges from 0.3022 s to 0.0473 s for the first 50 modes. The natural periods are listed in Table 3.2. The fundamental period is 0.1962 s.

Table 3.2 Natural Periods and Mass Participation Factors

Mode	Period (s)	Description	Mass Participation Factors					
			UX	UY	UZ	RX	RY	RZ
1	0.3022		0.000	**0.301**	0.000	0.699	0.000	0.230
2	0.1962	UX 1	**0.202**	0.000	**0.243**	0.000	0.170	0.000
3	0.1866		0.000	0.000	0.000	0.000	0.000	0.000
4	0.1864		0.000	0.000	0.000	0.000	0.000	0.000
5	0.1852	UY 1	0.000	**0.132**	0.000	0.006	0.000	0.106
6	0.1850		0.044	0.000	0.010	0.000	0.006	0.000
7	0.1697		0.000	0.000	0.000	0.000	0.000	0.000
8	0.1695		0.000	0.000	0.005	0.000	0.006	0.000
9	0.1647		0.000	0.000	0.001	0.000	0.001	0.000
10	0.1645		0.001	0.000	0.004	0.000	0.005	0.000
11	0.1413		0.015	0.000	0.057	0.000	0.034	0.000
12	0.1406		0.000	0.000	0.001	0.000	0.003	0.000
13	0.1399		0.005	0.000	0.016	0.000	0.012	0.000
14	0.1388		0.000	0.000	0.000	0.000	0.000	0.000
15	0.1384		0.000	0.000	0.001	0.000	0.001	0.000
16	0.1379		0.000	0.000	0.004	0.000	0.001	0.000
17	0.1378		0.001	0.000	0.002	0.000	0.004	0.000
18	0.1373		0.000	0.000	0.000	0.003	0.000	0.000
19	0.1353		0.000	0.000	0.000	0.004	0.000	0.000
20	0.1240	UZ 1	0.058	0.000	**0.434**	0.000	0.291	0.000
21	0.1184		0.000	0.001	0.000	0.000	0.000	0.032
22	0.1155		0.001	0.000	0.002	0.000	0.004	0.000
23	0.1149		0.001	0.000	0.019	0.000	0.016	0.000
24	0.1109		0.000	0.000	0.001	0.000	0.005	0.000
25	0.1108		0.000	0.000	0.000	0.000	0.000	0.000
26	0.0910		0.000	0.000	0.000	0.000	0.000	0.000
27	0.0907		0.002	0.000	0.000	0.000	0.000	0.000
28	0.0892		0.002	0.000	0.000	0.000	0.000	0.000
29	0.0891		0.000	0.000	0.000	0.000	0.000	0.000
30	0.0874	RX 1	0.000	**0.008**	0.000	**0.189**	0.000	0.009
31	0.0741		0.000	0.000	0.000	0.000	0.000	0.000
32	0.0738		0.000	0.000	0.000	0.000	0.000	0.000
33	0.0737		0.000	0.000	0.000	0.000	0.000	0.000
34	0.0723		0.000	0.000	0.000	0.006	0.000	0.000
35	0.0716	UZ 2	0.000	0.000	**0.001**	0.000	0.114	0.000
36	0.0702		0.000	0.000	0.000	0.000	0.000	0.000
37	0.0698		0.000	0.000	0.000	0.000	0.027	0.000
38	0.0674		0.000	0.000	0.000	0.000	0.000	0.000

Table 3.2 Continued Natural Periods and Mass Participation Factors

Mode	Period (s)	Description	Mass Participation Factors					
			UX	UY	UZ	RX	RY	RZ
39	0.0674		0.000	0.000	0.000	0.000	0.000	0.000
40	0.0650		0.000	0.000	0.000	0.000	0.000	0.041
41	0.0620		0.001	0.000	0.000	0.000	0.000	0.000
42	0.0605		0.000	0.000	0.000	0.000	0.000	0.000
43	0.0598		0.000	0.000	0.000	0.000	0.000	0.000
44	0.0597		0.000	0.000	0.000	0.000	0.000	0.000
45	0.0562		0.000	0.000	0.000	0.000	0.000	0.000
46	0.0562		0.000	0.000	0.000	0.000	0.000	0.000
47	0.0534	UZ 3	0.000	0.000	**0.100**	0.000	0.074	0.000
48	0.0499		0.000	0.000	0.000	0.000	0.000	0.000
49	0.0498		0.000	0.000	0.000	0.000	0.000	0.000
50	0.0473	RX 2	0.000	0.000	0.000	**0.007**	0.000	0.000

Figures 3.8 (a), (b) and (c) show the first mode shapes in isometric, elevation and plan views respectively. The natural period of this mode is 0.3022 s. Based on Figure 3.8 (a), (b) and (c) the vibration is mainly in the superstructure; however, the deck is stable for this case. So the mode 1 isn't the first mode of transverse even if mass participation factor is 0.301 in transverse direction.

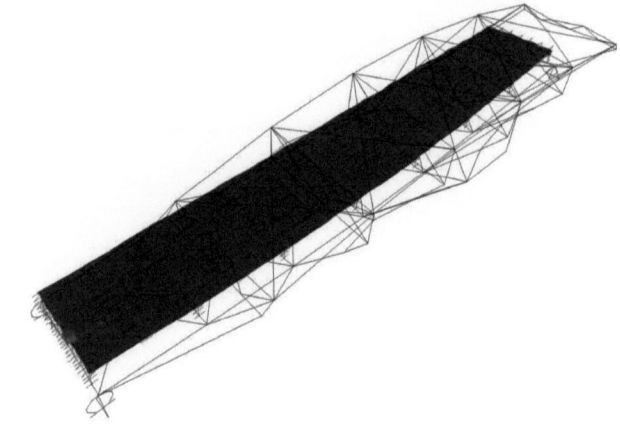

(a) Isometric View

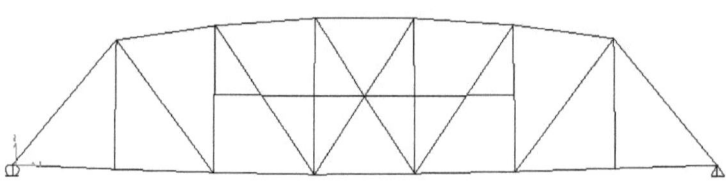

(b) Elevation View

Figure 3.8 Mode Shape Corresponding to the First Natural Period

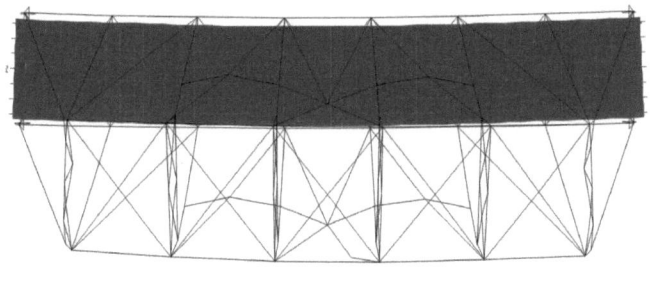

(c) Plan View

Figure 3.8 Continued Mode Shape Corresponding to the First Natural Period

The second mode, with a period of 0.1962 s, is shown in Figures 3.9 (a), (b) and (c). The mass participation for this mode is 0.202 in longitude direction and 0.243 in vertical direction. Compared with Figure 3.9 (a), (b), (c) and mass participation factor, second mode is the first mode in longitudinal direction.

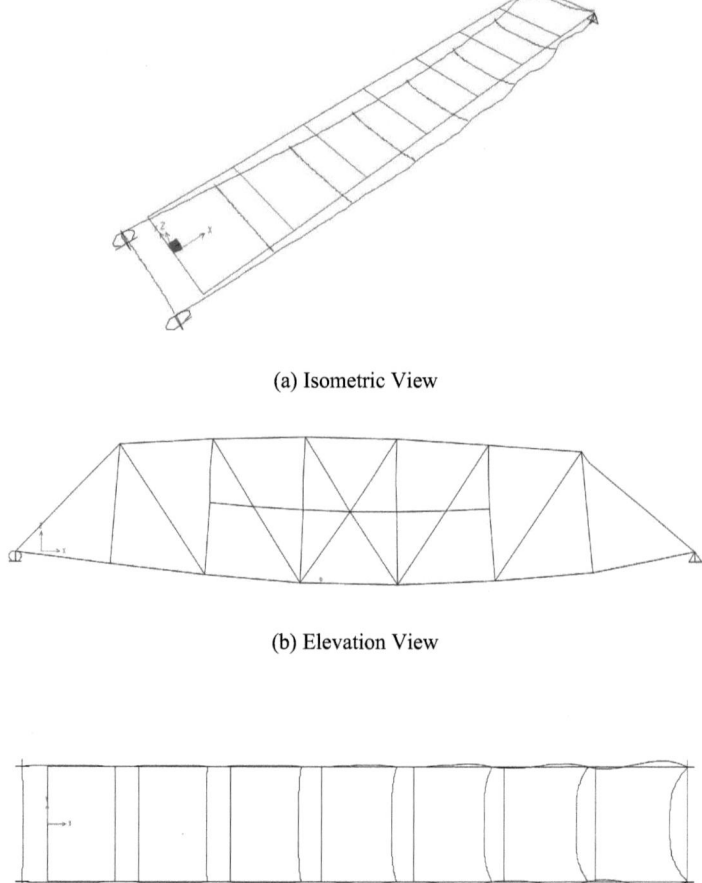

(a) Isometric View

(b) Elevation View

(c) Plan View

Figure 3.9 Mode Shape Corresponding to the Second Natural Period

(1st Longitudinal Mode Shape, 0.1962 s)

Figures 3.10 (a), (b) and (c) show the fifth mode shape with a period of 0.1962 s. The mass participation factor in transverse direction is 0.132. Based on Figures 3.10 (a), (b) and (c), this mode is the first mode in transverse direction.

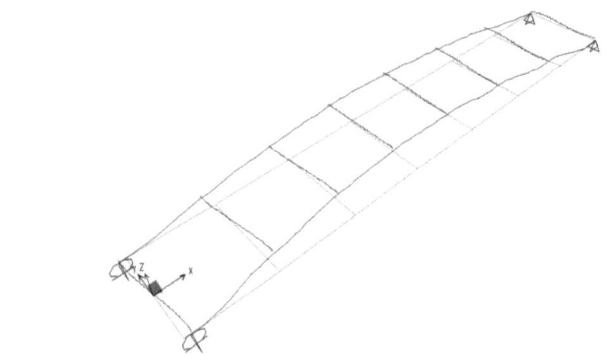

(a) Isometric View

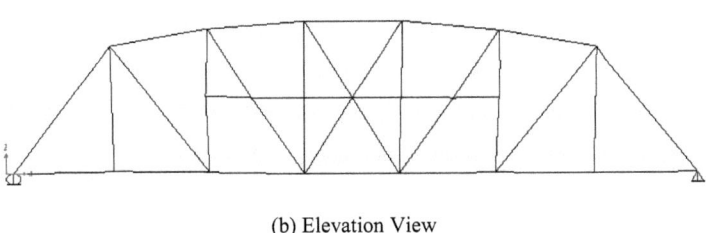

(b) Elevation View

Figure 3.10 Mode Shape Corresponding to the Fifth Natural Period

(1st Transverse Mode Shape, 0.1852 s)

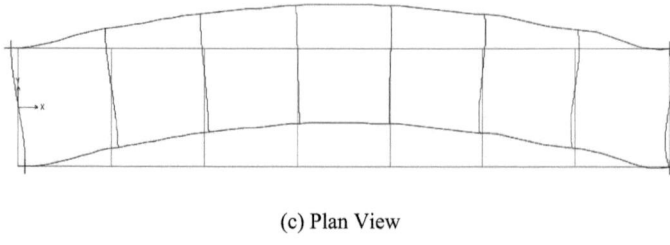

(c) Plan View

Figure 3.10 Continued Mode Shape Corresponding to the Fifth Natural Period

(1st Transverse Mode Shape, 0.1852 s)

Figures 3.11 (a), (b) and (c) show the twentieth mode shape with a period of 0.1240 s. The mass participation factor for this mode is 0.434 in vertical direction. This is the first mode in vertical direction based on figures and mass participation factor.

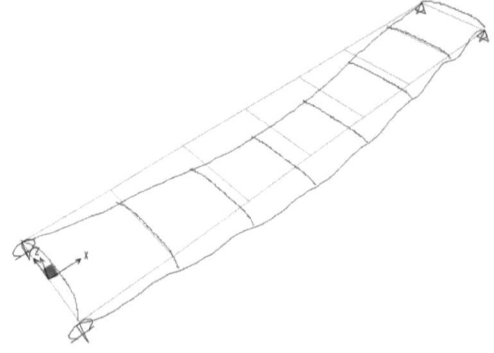

(a) Isometric View

Figure 3.11 Mode Shape Corresponding to the Twentieth Natural Period

(1st Vertical Mode Shape, 0.1240 s)

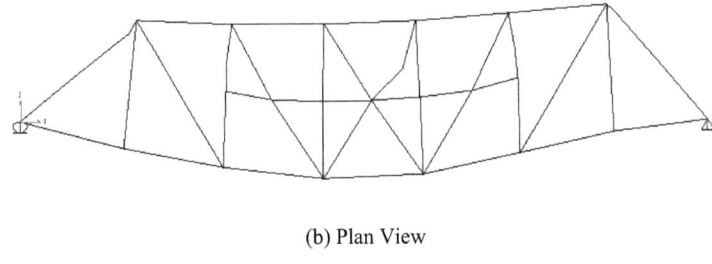

(b) Plan View

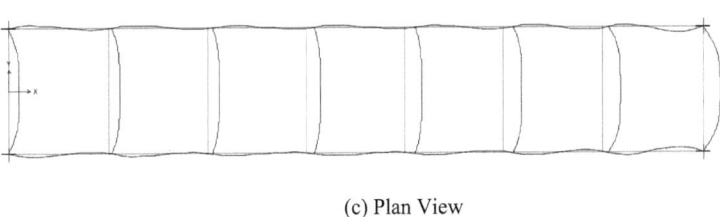

(c) Plan View

Figure 3.11 Continued Mode Shape Corresponding to the Twentieth Natural Period

(1st Vertical Mode Shape, 0.1240 s)

Figures 3.12 (a), (b) and (c) show the thirtieth mode shape with a period of 0.0874 s. The mass participation factor is 0.189 in torsional direction. This is the first mode of torsion.

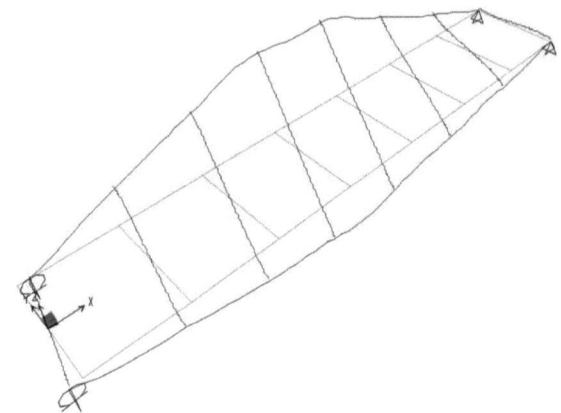

(a) Isometric View

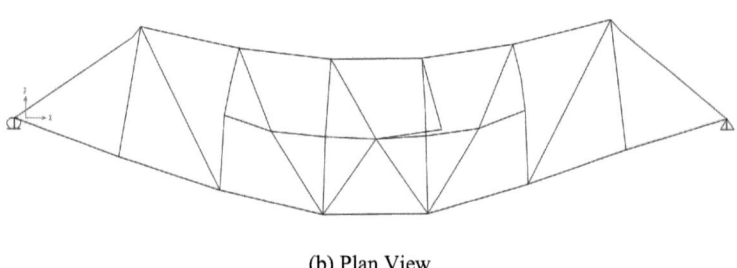

(b) Plan View

Figure 3.12 Mode Shape Corresponding to the Thirtieth Natural Period

(1st Torsional Mode Shape, 0.0874 s)

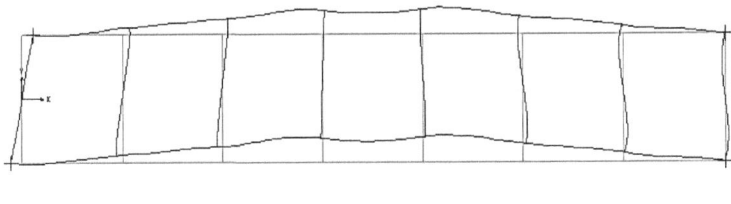

(c) Plan View

Figure 3.12 Continued Mode Shape Corresponding to the Thirtieth Natural Period

(1st Torsional Mode Shape, 0.0874 s)

Figures 3.13 (a), (b) and (c) show the thirty fifth mode shape with a period of 0.0716. The mass participation factor is 0.001 in vertical direction. Based on figures 3.13 (a), (b) and (c), this is the second mode in vertical direction even if it has very low mass participation factor.

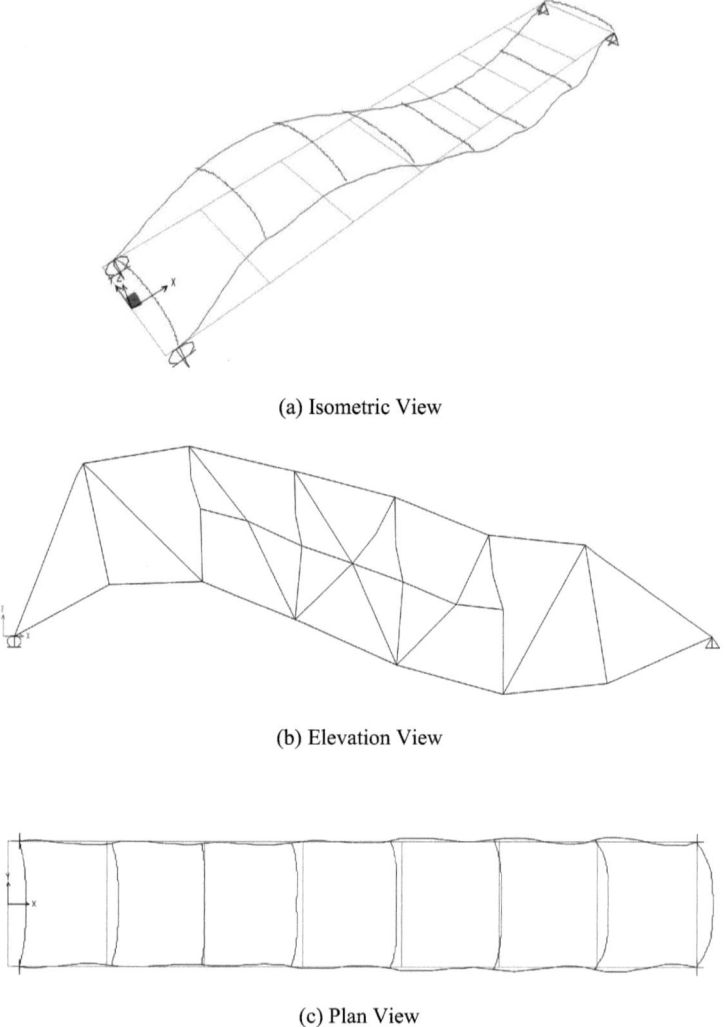

(a) Isometric View

(b) Elevation View

(c) Plan View

Figure 3.13 Mode Shape Corresponding to the Thirty-Fifth Natural Period

(2nd Vertical Mode Shape, 0.0716 s)

39

Figures 3.14 (a), (b) and (c) show the forty-seventh mode shape with a period of 0.0534 s. The mass

participation factor is 0.100 in vertical direction. Based on figures 3.14 (a), (b) and (c), this is the third

mode in vertical direction.

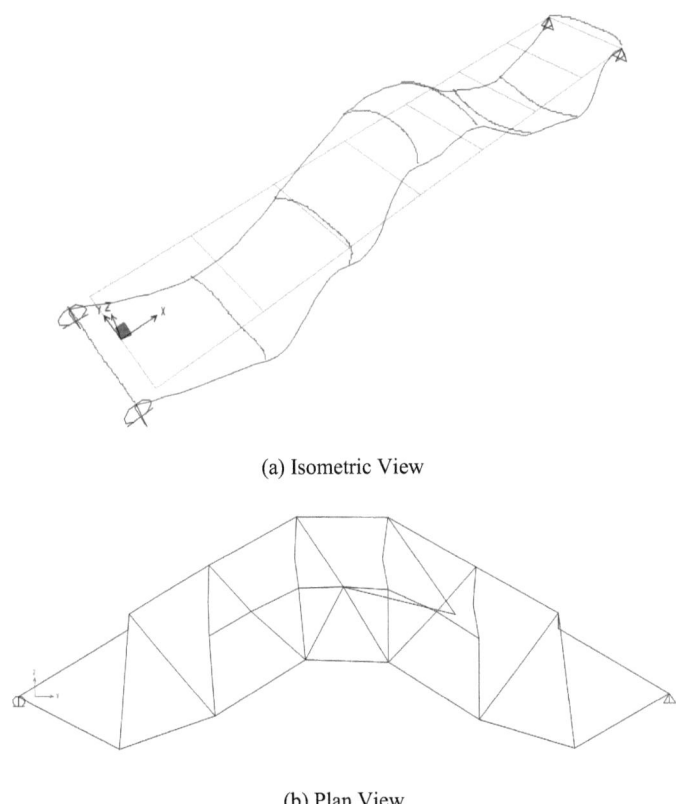

(a) Isometric View

(b) Plan View

Figure 3.14 Mode Shape Corresponding to the Forty-Seventh Natural Period

(3rd Vertical Mode Shape, 0.0534 s)

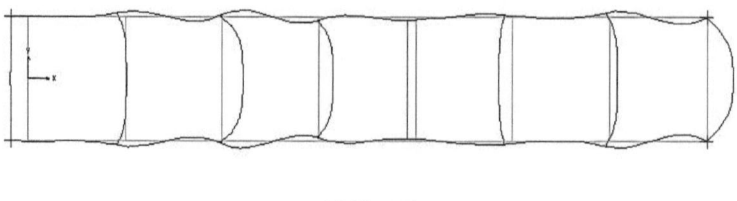

(c) Plan View

Figure 3.14 Continued Mode Shape Corresponding to the Forty-Seventh Natural Period

(3rd Vertical Mode Shape, 0.0534 s)

Figures 3.15 (a), (b) and (c) show the forty-seventh mode shape with a period of 0.0473 s. The mass participation factor is 0.007 in torsional direction. Based on figures 3.15 (a), (b) and (c), this is the second mode in torsional direction.

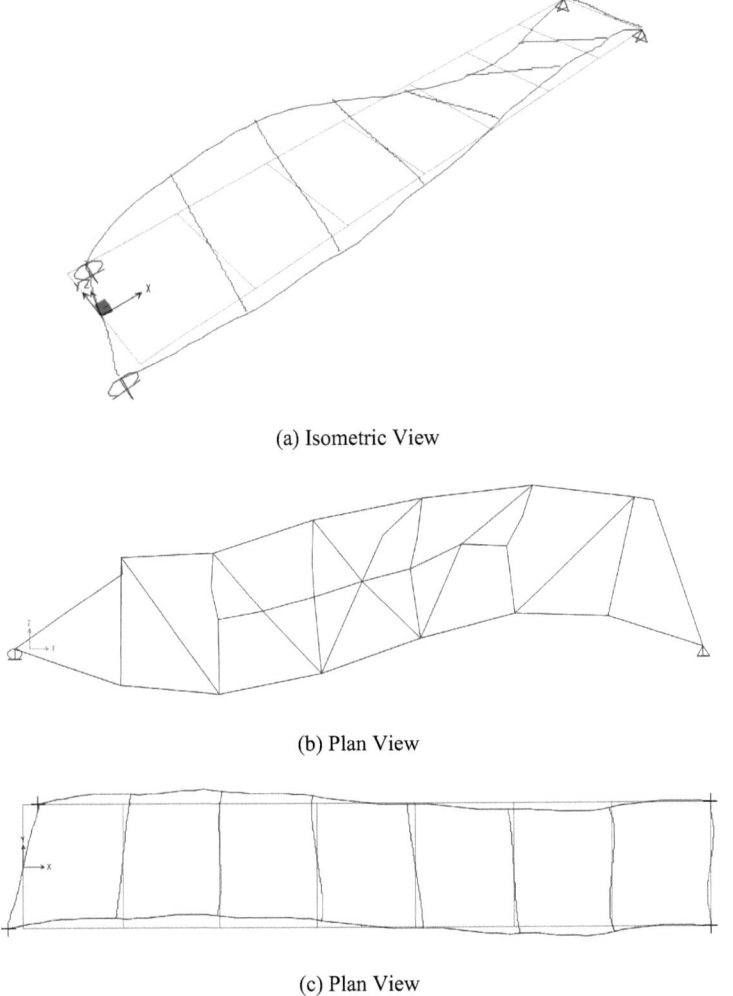

(a) Isometric View

(b) Plan View

(c) Plan View

Figure 3.15 Mode Shape Corresponding to the Fiftieth Natural Period

(2nd Torsional Mode Shape, 0.0473 s)

The accelerometer sensor plan follows standard procedures for acquisition of dynamic properties (or signature) of the structure. Lower modes and corresponding frequencies were planned to be measured by accelerometers. From the modal analysis, lower natural periods and mode shapes for four directions have been successfully identified (Fig. 3.16). Because of the limited number of accelerometers, accelerometers should be fixed at the best positions to measure the first three modes and corresponding frequencies. Finite element modal analysis predicted the mode shapes. That gives a guideline for the placement of accelerometers.

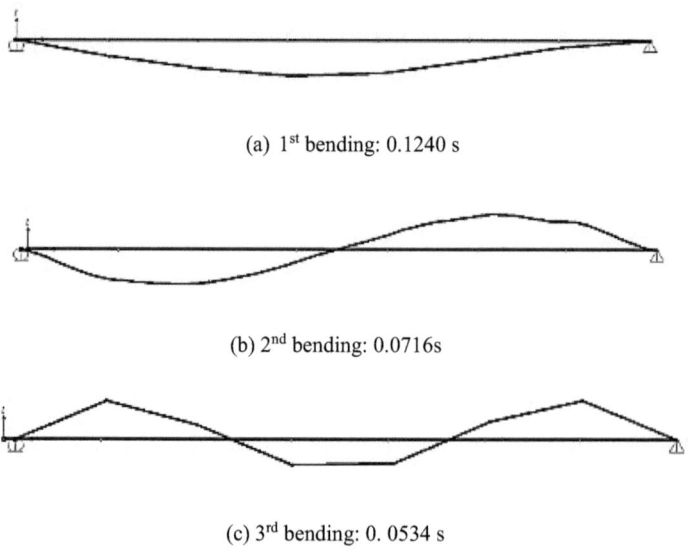

(a) 1st bending: 0.1240 s

(b) 2nd bending: 0.0716s

(c) 3rd bending: 0. 0534 s

Figure 3.16 Mode Shapes and Natural Periods

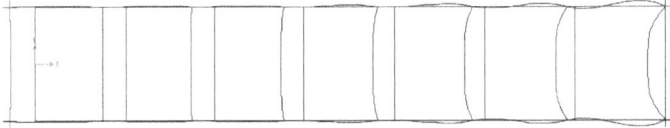

(d) 1st longitudinal: 0.1962 s

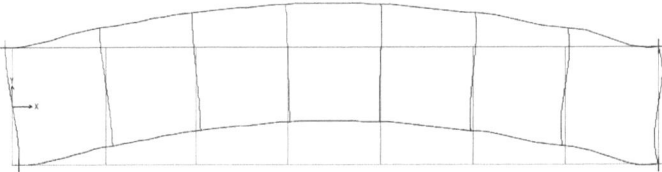

(e) 1st transverse: 0.1852 s

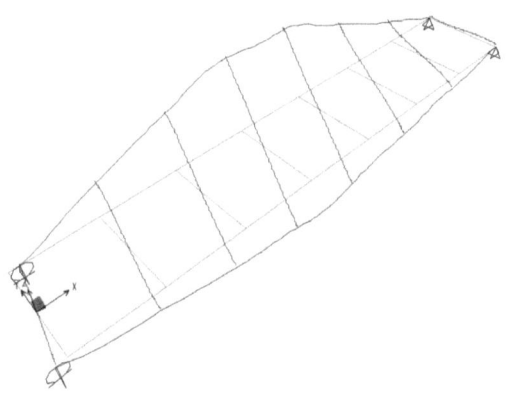

(f) 1st torsion: 0.0874s

Figure 3.16 Continued Mode Shapes and Natural Periods

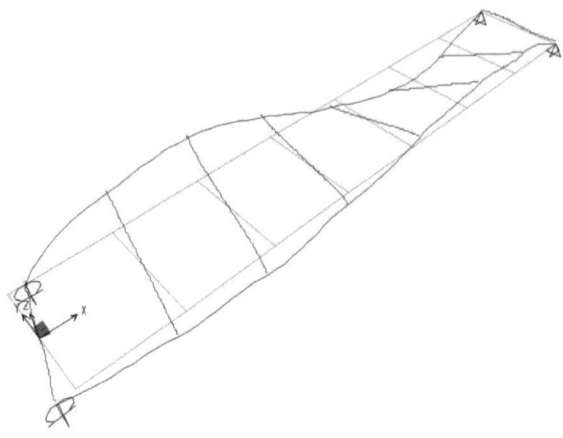

(g) 2nd torsion: 0.0473 s

Figure 3.16 Continued Mode Shapes and Natural Periods

According to modal analysis result, accelerometers were placed at the bridge deck level (bottom chords of the trusses) along the length of the bridge (see Fig. 3.17) to measure the natural frequencies and mode shapes of the bridge structure. This information can also be used for monitoring the global condition of the bridge and for mode identification.

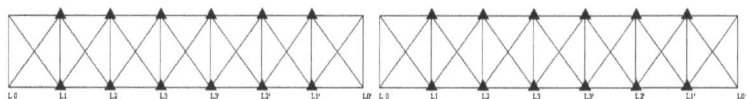

Figure 3.17 Preliminary Accelerometer Sensor Layout

3.3 Local Finite Element Analysis

Steel portal frames were designed assuming that beam-to-column joints are ideally pinned or fully rigid. This simplified the analysis and structural design processes, but at the expense of not obtaining a detailed understanding of the behavior of the joints which are semi-rigid in reality. In frame analysis, joint rotational behavior should be considered. This is usually done by using the moment-rotation curve. In this research, a local finite element model was built to determine rotational stiffness of selected section.

Moving load analysis revealed that there are considerable amounts of bending moment in girder-to-column section (Fig. 3.18). Therefore it is essential to determine the rotational stiffness of girder-to-column and update global modal's local stiffness.

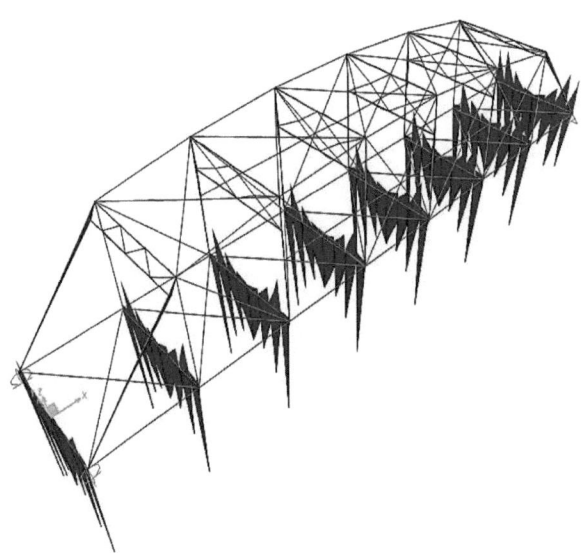

Figure 3.18 Moment Diagram

Girder-to-column (see Fig. 3.19) was made up of angle cleats riveted to the flange of the members. A local riveted bridge connection model was developed using ABAQUS. The refined connection model, which consists of an assembly of a lower chord truss and a vertical column modeled as fixed. The remaining girder was modeled as a cantilever. The connection consists of four angles, each rived to the girder web and column flange. All of the elements were modeled by using 8-noded brick elements with full integration. A Young's modulus of 200 GPa, Poisson's ratio of 0.3, and linear elastic behavior were assumed for the finite element analysis. Two point loads were applied on the end of girder which stands for the moment. And part of the girder was assumed as rigid body to reduce the influence of the girder bending. So the rotation was totally caused by connection. Force was increased in steps in order to investigate the moment-rotation behavior of the connection.

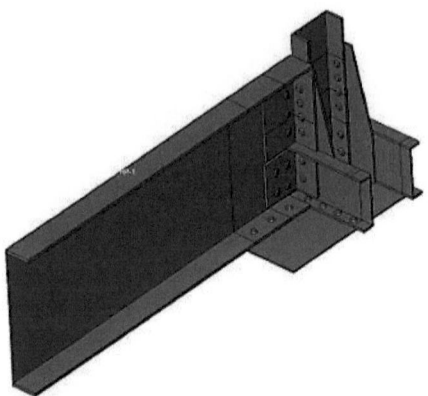

Figure 3.19 Girder-to-Column Connection

The moment-rotation behavior of the connection is shown in Fig. 3.20. The effect of friction was taken into account in the finite element model by defining a coefficient of friction of 0.3 between the surfaces in contact. Then the moment can transfer to this connection by updating general modal's rotational stiffness. In that case, the connection's behavior can be identified when different kinds of vehicles cross the bridge.

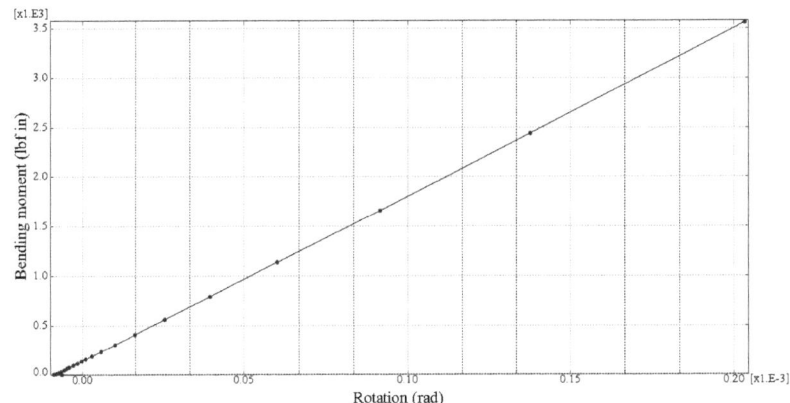

Figure 3.20 Moment-Rotation L1 Connection

The finite element local model may be used to evaluate the L1 connection's rotational stiffness. The hot-spot strain picture can be obtained from the finite element analysis results (Fig. 3.21).

Figure 3.21 L1 Hot Spot Strain

Hot-spot strain shows there is no obvious strain in the outside gusset plate. The fatigue damage should begin in the inside gusset plate. From the fracture critical bridge inspection report from 2007 to 2010, there was no crack observed in the inside gusset plate. So cracks at the outside gusset plate were due to other types of load damage. Those kinds of cracks can be imitated by local finite element model which can figure out how serious those cracks influence on the deformation of this connection.

3.4 Crack Gage

The crack gauges will show movement and progression of cracking at the sensor locations. Crack gauges are also able to track the number of loading cycles for establishing remaining service life.

According to QA Services, Inc. 2011 report, there were 21 cracks have been identified as locations NDE 1 through NDE 19 (Fig. 3.22) [14]. In 2012 field inspection, a new crack was found identified as locations as NDE 20 which was selected to monitor the crack propagation.

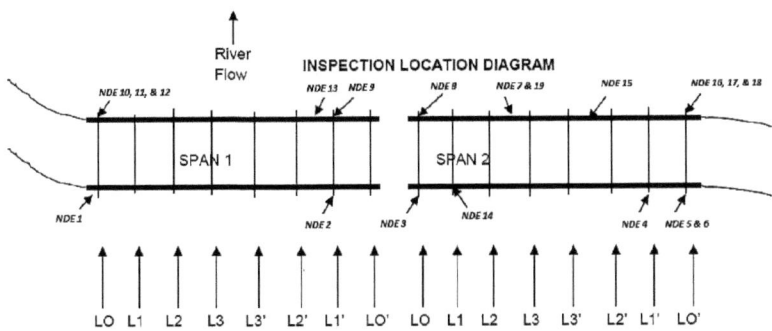

Figure 3.22 Inspection Location Diagram (Photo Courtesy of ADOT&PF)

Cracks at the end of lower chord's lower flange were caused by fatigue damage of tension. For each span's every end, NDE 6 a (Fig. 3.23), NDE 6 c (Fig. 3.24) and NDE 11 b (Fig. 3.25) were selected to be monitoring. Those cracks have high potential to propagation.

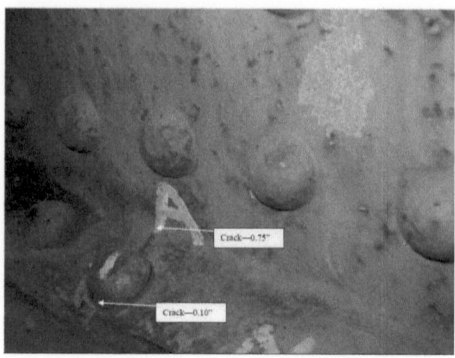

Figure 3.23 Cracked Interior and Exterior Channels (NDE 6a)

(Photo Courtesy of ADOT&PF)

Figure 3.24 Cracked Interior and Exterior Channels (NDE 6c)

(Photo Courtesy of ADOT&PF)

Figure 3.25 Cracked Lower Flange of Exterior Channel Lower Chord (NDE 11b)

(Photo Courtesy of ADOT&PF)

A large tear with distortion was noted at upstream U1'-L1', Span 1. The 1-1/2'' tear is in outside gusset plate connections to the lower chord at L1' (see Fig 3.26). There is a similar damage at the downstream with a 3''×5'' tear (Fig. 3.27). The adjacent steel has indentation from cables bearing against the plates, suggesting that the tear was due to erection damage [15].

Figure 3.26 Torn Gusset Plate (NDE 2)

Figure 3.27 Torn Gusset Plate (NDE 9)

NDE 8c was selected to monitor. Cracked gusset plate extends from bottom of gusset to under rivet head. Magnetic particle indication reveals through crack (Fig. 3.28). This is the exterior gusset plate which contains the weld repair identified in location 8 [15].

Figure 3.28 Cracked Gusset Plate around Rivet Head (NDE 8c)

The end-span gusset plants (Fig. 3.29, Fig. 3.30) were damaged because of the replacement of this bridge. The workers cut those gusset plants for relocation and filleted it after settled down. NED 5 (Fig. 3.30) was selected to monitor based on the dimension of the crack.

Figure 3.29 Gusset Plate Welded Repair (NDE 3)

Figure 3.30 Gusset Plate Welded Repair (NDE 5)

From field inspection, the research team divided cracks into three kinds: cracks at end of lower chord's lower flanges, cracks at the mid-span outside gusset plants, and weld repairs the end-span's gusset plants (Fig. 3.31). A preliminary crack gage layout is shown in Fig. 3.32.

a. Crack at End of Lower Chord Lower Flanges (NDE 6)

Figure 3.31 Different Cracks

b. Crack at the Mid-span Outside Gusset Plants (NDE 2)

c. Weld Repair at the End-span Gusset Plants (NDE 8)

Figure 3.31 Continued Different Cracks

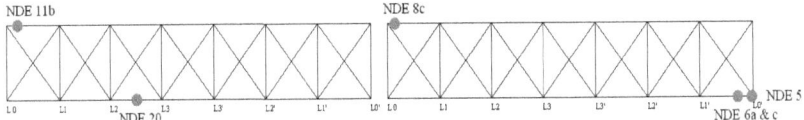

Figure 3.32 Preliminary Crack Gage Layout

3.5 Preliminary Sensor Layout

A preliminary sensor layout (including accelerometers, strain and temperature sensors, crack gauges, etc.) is shown in Fig. 3.33. It was proposed that a total of 56 sensors will be installed for monitoring of this bridge. However, since this study is aimed to monitor gradual degradation of the bridge, the sensor arrangement does not cover all the cracks but provides information about changes in the load path when cracks gradually increase in length. The design of the bridge structure allows for the use of a minimal number of temperature compensation sensors. In this case a total of four temperature sensors were separated in each truss. Preliminary structural analysis showed that the diagonal members of the trusses are fracture critical members. For this reason a strain sensor should be placed to monitor of these members resulting in a total of eight sensors. As the main load path the lower chord members should also be monitored, especially those weld repaired lower chords. Strain sensors were located near the middle points of each truss for an additional four sensors and one sensor for weld repair and lower chord truss. Sixteen strain sensors were allocated for the monitoring of the top chords of each truss.

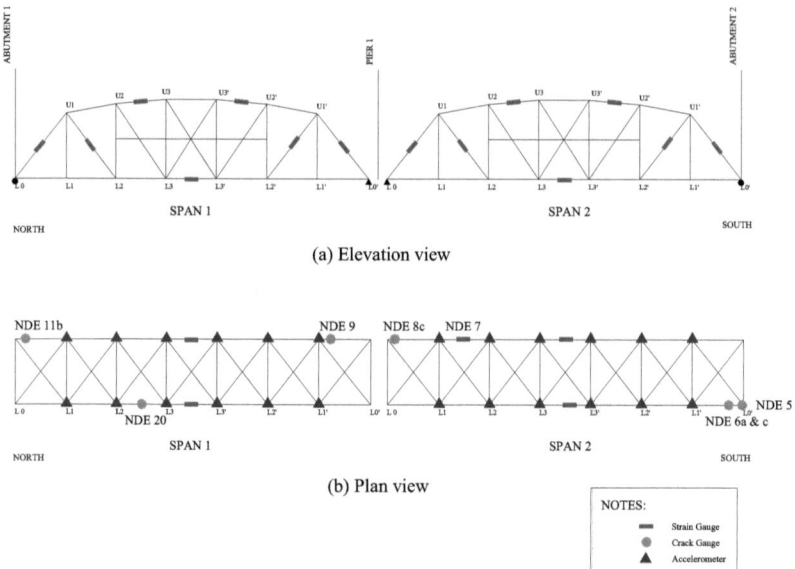

Figure 3.33 Preliminary Sensor Layout on Trusses

Because of the poor conditions around the bridge supports, including oxidation and soil build up. The expansion supports will to be monitored for rotation with tilt meters. If all supports are not free to rotate as they should the bridge may exhibit a twisting condition. Finally, an additional seven crack sensors were located near specific defects in gusset plates and channel flanges to monitor for crack activity. The following table is a brief summary of number and locations of the sensors.

Table 3.3 Summary Number of Sensors

Sensor and locations	Number of Sensors
Strain Sensors on the Top Chord Members	16
Strain Sensors on the Diagonal Members	8
Strain Sensors on the Lower Chord Members	5
Crack Sensors	7
Portable Accelerometers	12
Tilt meter (at expansion supports)	4
Temperature Sensors	4
Total	56

The objective of this sensor plan was to utilize the optimized number and types of sensors to monitor structural health and to develop an understanding of the primary causes for damage. Final placement of the sensors may slightly move due to physical space constrictions.

3.6 Types of Monitoring

Dynamic monitoring: The accelerometer sensor plan follows standard procedures for acquisition of dynamic properties of the structure. Accelerometers were placed at the bridge deck level (lower chords of the trusses) along the length of the bridge to provide the natural periods and mode shapes of the bridge structure. This information can be used for monitoring the global condition of the bridge. It can also be used to calibrate and validate structural analysis models. A more accurate computer model allows for more confidence in structural evaluation and future analysis for repair or design.

Stress monitoring: ADOT&PF's annual fracture critical inspections of the Klehini River Bridge have found torn gusset plates, and cracked rivet holes on the primary trusses, as well as damaged sway and lateral bracing members. Strain gauges were placed at selected truss members. The strain gauges can provide a stress history of the members to assess if the members are being over stressed.

Deformation/crack monitoring: The crack gauges can show movement and progression of cracking at the sensor locations. Crack gauges are also able to track the number of loading cycles for establishing remaining service life. Analysis of both the strain gauge and crack gauge data will be used in ascertaining the cause of the cracks at rivet holes.

3.7 Equipment

The fiber optic sensors can be connected in series. Fusion splices are preferred in order to minimize loss. Armored cable, cable in conduit, or other similar type of protection keeps the sensors from weather exposure. The optical fiber sensor data is carried through optical leads and routed to the optical interrogator unit at the site via a multiplexer (Fig. 3.34). Optical data is converted to electrical signals at the interrogator and the data is fed into the local computer (the controller & data acquisition module). Data from the local computer is transmitted to the internet via satellite since hard wire internet is not available at the site.

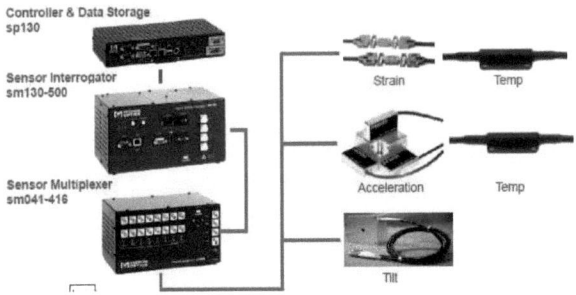

Figure 3.34 System Configuration (Photo Courtesy of Dr. Yongtao Dong)

The optical system will be housed inside a NEMA enclosure with controlled temperature and humidity (Fig. 3.35). The required conditions and the necessary temperature and humidity controls will be explored before a final system is chosen. The NEMA enclosure should be NEMA 4 or 4X rating with interior insulation, door operated light fixture, heater, and fan with thermostat controls. The enclosure have at least 5 openings for electrical, internet (satellite or DSL), and fiber connections. A disconnect and fuse block is also needed. An approximately 8 ft^3 interior space is required to host the optical system.

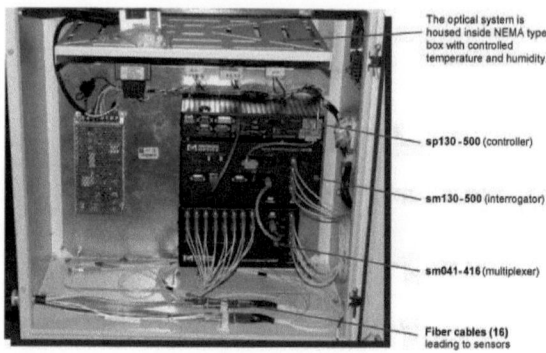

Figure 3.35 SHM System in NEMA Enclosure (Photo Courtesy of Dr. Yongtao Dong)

3.5 Power Supply and Internet for Remote Monitoring

The power supply for the SHM system is from a utility pole near the bridge site through a transformer. There is an active power line at the site and a pole with a meter is installed at the Klehini River Bridge Kleihny. The research team verified that sufficient power is available to run the SHM and power the sensors. At other remote sites, power supply equipment such as batteries, charger controllers, wind turbines and solar are possible choices.

Since there is no cell service available at the Klehini Bridge crossing, the SHM system needs to be integrated to the internet for remote monitoring. There is a land telephone line which crosses the bridge and is operated by Alaska Power and Telephone (AP&T). Per our conversation with AP&T, a DSL internet service with 4 Mbps speed through the phone line. (Currently the fastest speed at the bridge is 512 Kbps).

3.6 Installation of SHM and Integration of the System

A standard and scope of installation work will be developed for a telecom contractor to install the fiber optic sensors on the bridges. Armored cable, cable in conduit, or other similar type of protection keeps the sensors from weather exposure. The optical fiber sensor data is carried through optical leads and routed to the optical interrogator unit at the site. Optical data is converted to electrical signals at the interrogator and the data is fed into the local computer (the controller & data acquisition module). Data from the local computer is transmitted to remote computer via DSL Internet.

Chapter 4 Conclusions and Future Work

This study has presented the design method for developing a structural health monitoring system. The Fiber Optic Sensor system was selected for the harsh cold conditions. A global finite element model was built based on as-built condition by using SAP2000. Moving load analysis following American Association of State Highway and Transportation Officials (AASHTO) 2007 was used to figure out critical members. Strain gages were placed on those critical members to ensure the live load won't out of design limit in actual traffic conditions. From the modal analysis, the lowest mode shapes and natural periods in vertical, transverse, longitudinal and rotational direction were found based on the mass participation factor. The mode shapes indicated the best position to place the accelerometers. After field measurement of mode shapes and natural periods, the field dynamic results will be calibrated with the finite element results which can identify the accuracy of the finite element model. A local finite model was developed by ABAQUS to figure out the rotational stiffness of one connection. The preliminary layout of crack gages was based on recent inspection reports and the field inspections. Cracks were classified into three kinds: cracks at the end of lower chord lower flanges, cracks at the mid-span outside gusset plants, and weld repair at the end-span gusset plants. Cracks were selected for monitoring depended on the possibilities of expansion.

Based on the recent inspection reports of the bridge provided by ADOT&PF, a second finite element global model including the detected degradation/defects on the bridge will be created to relate the "current in use" condition by using ABAQUS. This modified model will be calibrated from field static and dynamic testing to represent the real bridge condition and be used to predict the response of the bridge in active traffic loading. Several local finite element models will be built to simulate the cracks and the semi-rigid connection. The local finite element model can be connected with the global model by using reference points which enhance the accuracy of the finite element model. Mode identification will be performed by comparing the numerical dynamic results with field measurement. The finite element model will be modified in the future based on model identification's result.

Finally, the modified finite element model can help researchers check local damage's influence on global behaviors and influence of the different kinds of traffic loads on local damage, which is essential information to predict the future behavior of this bridge. With accurate finite element model, the load rating can be conducted based on the guidance of the Manual for Bridge Evaluation. The load rating results can show each bridge members' condition and give bridge owner guidance for repair or replacement of this bridge.

References

(1) Alaska DOT&PF, Fracture Critical Inspection Report - Bridge Number 1216: Klehni River Bridge, 2008

(2) Alaska DOT&PF, Fracture Critical Inspection Report - Bridge Number 1216: Klehni River Bridge, 2010

(3) Karbhari, V. M. and Ansari F. "Structural Health Monitoring of Civil Infrastructure Systems", publications in the European Workshop on Structural Health Monitoring, 5th edition, 2010

(4) Phares, B. M., Wipf, T. J., Greimann, L. F. and Lee Y., Health monitoring of Bridge Structures and components using smart Structure Technology", Vol. 1&2, Iowa State University, 2005

(5) Whelan, M.J., Fuchs, M.P., Gangone, M.V. and K.D. Janoyan, "Development of a Wireless Bridge Monitoring System for Condition Assessment Using Hybrid Techniques", Clarkson University, 2008

(6) Whelan, M.J. and K.D. Janoyan, "Wireless Network for Monitoring of Geo-Structural Systems", Clarkson University, 2009

(7) Stein, P, "Utilization of handheld field testing system for improvements of bridge load rating values in PONTIS, Iowa State University, 2005

(8) Hemphill, D, "Structural health monitoring system for the east 12th bridge", Iowa State University, 2004

(9) Miyashita, T and Nagai, M. " Development of sensor node and analytical framework for vibration-based structural health monitoring of existing bridges," 5th World Conf. on Structural Control & Monitoring, Tokyo, 2010

(10) Dong, Y., Liu, H. and Song R., "Bridge structural health monitoring and deterioration detection – synthesis of knowledge and technology", Report INE/AUTC 11.xx, University of Alaska Fairbanks, 2011

(11) Rytter, A, 'Vibration based inspection of civil engineering structures.' PhD Thesis, Department of Building Technology and Structural Engineering, Aalborg University, Denmark, 1993

(12) Grivas, D. A. and Garlock, M. Sensing systems for bridges: an assessment of the state-of-the-art. In: Mahmoud, K. M. (Ed) Proceedings of the Second New York City Bridge Conference. New York, NY, USA: A.A. Balkema. 2003. 269-284.

(13) Pines, D., and Aktan, A.E. 2002. Status of SHM of long-span bridges in the United States. Progress in Structural Engineering and Materials, Vol. 4, No.4, pp. 372-380.

(14) QA Services, Inc, On System Bridge Inspection – Bridge Number 1216: Klehini River Bridge, 2011

(15) Alaska DOT&PF, Fracture Critical Inspection Report - Bridge Number 1216: Klehni River Bridge, 2007

Appendix

Table A.1 Mass Participation Factors

Mode	Period	Mass Participation Factor					
		UX	UY	UZ	RX	RY	RZ
1	0.3022	0.000	0.301	0.000	0.699	0.000	0.230
2	0.1962	0.202	0.000	0.243	0.000	0.170	0.000
3	0.1866	0.000	0.000	0.000	0.000	0.000	0.000
4	0.1864	0.000	0.000	0.000	0.000	0.000	0.000
5	0.1852	0.000	0.132	0.000	0.006	0.000	0.106
6	0.1850	0.044	0.000	0.010	0.000	0.006	0.000
7	0.1697	0.000	0.000	0.000	0.000	0.000	0.000
8	0.1695	0.000	0.000	0.005	0.000	0.006	0.000
9	0.1647	0.000	0.000	0.001	0.000	0.001	0.000
10	0.1645	0.001	0.000	0.004	0.000	0.005	0.000
11	0.1413	0.015	0.000	0.057	0.000	0.034	0.000
12	0.1406	0.000	0.000	0.001	0.000	0.003	0.000
13	0.1399	0.005	0.000	0.016	0.000	0.012	0.000
14	0.1388	0.000	0.000	0.000	0.000	0.000	0.000
15	0.1384	0.000	0.000	0.001	0.000	0.001	0.000
16	0.1379	0.000	0.000	0.004	0.000	0.001	0.000
17	0.1378	0.001	0.000	0.002	0.000	0.004	0.000
18	0.1373	0.000	0.000	0.000	0.003	0.000	0.000
19	0.1353	0.000	0.000	0.000	0.004	0.000	0.000
20	0.1240	0.058	0.000	0.434	0.000	0.291	0.000
21	0.1184	0.000	0.001	0.000	0.000	0.000	0.032
22	0.1155	0.001	0.000	0.002	0.000	0.004	0.000
23	0.1149	0.001	0.000	0.019	0.000	0.016	0.000
24	0.1109	0.000	0.000	0.001	0.000	0.005	0.000
25	0.1108	0.000	0.000	0.000	0.000	0.000	0.000
26	0.0910	0.000	0.000	0.000	0.000	0.000	0.000
27	0.0907	0.002	0.000	0.000	0.000	0.000	0.000
28	0.0892	0.002	0.000	0.000	0.000	0.000	0.000
29	0.0891	0.000	0.000	0.000	0.000	0.000	0.000
30	0.0874	0.000	0.008	0.000	0.189	0.000	0.009

Table A.1 Continued Mass Participation Factors

Mode	Period	Mass Participation Factor					
		UX	UY	UZ	RX	RY	RZ
31	0.0741	0.000	0.000	0.000	0.000	0.000	0.000
32	0.0738	0.000	0.000	0.000	0.000	0.000	0.000
33	0.0737	0.000	0.000	0.000	0.000	0.000	0.000
34	0.0723	0.000	0.000	0.000	0.006	0.000	0.000
35	0.0716	0.000	0.000	0.001	0.000	0.114	0.000
36	0.0702	0.000	0.000	0.000	0.000	0.000	0.000
37	0.0698	0.000	0.000	0.000	0.000	0.027	0.000
38	0.0674	0.000	0.000	0.000	0.000	0.000	0.000
39	0.0674	0.000	0.000	0.000	0.000	0.000	0.000
40	0.0650	0.000	0.000	0.000	0.000	0.000	0.041
41	0.0620	0.001	0.000	0.000	0.000	0.000	0.000
42	0.0605	0.000	0.000	0.000	0.000	0.000	0.000
43	0.0598	0.000	0.000	0.000	0.000	0.000	0.000
44	0.0597	0.000	0.000	0.000	0.000	0.000	0.000
45	0.0562	0.000	0.000	0.000	0.000	0.000	0.000
46	0.0562	0.000	0.000	0.000	0.000	0.000	0.000
47	0.0534	0.000	0.000	0.100	0.000	0.074	0.000
48	0.0499	0.000	0.000	0.000	0.000	0.000	0.000
49	0.0498	0.000	0.000	0.000	0.000	0.000	0.000
50	0.0473	0.000	0.000	0.000	0.007	0.000	0.000
51	0.0464	0.000	0.000	0.000	0.003	0.000	0.000
52	0.0436	0.009	0.000	0.000	0.000	0.011	0.000
53	0.0424	0.000	0.000	0.000	0.002	0.000	0.000
54	0.0416	0.000	0.000	0.000	0.000	0.000	0.000
55	0.0393	0.118	0.000	0.004	0.000	0.063	0.000
56	0.0381	0.000	0.019	0.000	0.006	0.000	0.014
57	0.0375	0.000	0.000	0.024	0.000	0.021	0.000
58	0.0351	0.000	0.000	0.000	0.000	0.000	0.001
59	0.0342	0.000	0.000	0.000	0.002	0.000	0.001
60	0.0338	0.000	0.005	0.000	0.019	0.000	0.004
61	0.0334	0.001	0.000	0.000	0.000	0.008	0.000
62	0.0325	0.000	0.000	0.000	0.000	0.000	0.000
63	0.0322	0.000	0.000	0.000	0.000	0.000	0.000

Table A.1 Continued Mass Participation Factors

Mode	Period	Mass Participation Factor					
		UX	UY	UZ	RX	RY	RZ
64	0.0321	0.000	0.000	0.000	0.000	0.000	0.000
65	0.0315	0.000	0.000	0.007	0.000	0.005	0.000
66	0.0309	0.000	0.000	0.000	0.001	0.000	0.000
67	0.0306	0.000	0.000	0.000	0.000	0.000	0.000
68	0.0306	0.000	0.000	0.000	0.000	0.000	0.000
69	0.0305	0.000	0.000	0.000	0.006	0.000	0.000
70	0.0301	0.000	0.000	0.000	0.000	0.000	0.000
71	0.0300	0.000	0.000	0.000	0.001	0.000	0.001
72	0.0298	0.000	0.000	0.000	0.000	0.000	0.000
73	0.0296	0.023	0.000	0.000	0.000	0.007	0.000
74	0.0282	0.000	0.000	0.000	0.003	0.000	0.000
75	0.0278	0.000	0.000	0.000	0.000	0.000	0.000
76	0.0274	0.000	0.000	0.000	0.000	0.000	0.000
77	0.0270	0.000	0.000	0.000	0.000	0.000	0.000
78	0.0268	0.000	0.000	0.000	0.000	0.000	0.000
79	0.0260	0.000	0.000	0.000	0.000	0.000	0.000
80	0.0256	0.009	0.000	0.000	0.000	0.003	0.000
81	0.0255	0.000	0.000	0.000	0.000	0.000	0.004
82	0.0244	0.000	0.000	0.000	0.000	0.000	0.000
83	0.0239	0.000	0.000	0.000	0.000	0.000	0.000
84	0.0236	0.000	0.000	0.000	0.000	0.000	0.000
85	0.0231	0.003	0.000	0.000	0.000	0.001	0.000
86	0.0226	0.004	0.000	0.000	0.000	0.004	0.000
87	0.0220	0.000	0.000	0.003	0.000	0.002	0.000
88	0.0211	0.001	0.000	0.000	0.000	0.001	0.000
89	0.0205	0.002	0.000	0.002	0.000	0.006	0.000
90	0.0202	0.000	0.000	0.000	0.000	0.002	0.000
91	0.0196	0.000	0.000	0.001	0.000	0.001	0.000
92	0.0191	0.000	0.000	0.003	0.000	0.003	0.000
93	0.0184	0.000	0.000	0.001	0.000	0.000	0.000
94	0.0183	0.000	0.000	0.000	0.000	0.003	0.000
95	0.0166	0.000	0.000	0.012	0.000	0.004	0.000
96	0.0165	0.001	0.000	0.005	0.000	0.025	0.000

Table A.1 Continued Mass Participation Factors

Mode	Period	Mass Participation Factor					
		UX	UY	UZ	RX	RY	RZ
97	0.0157	0.002	0.000	0.003	0.000	0.001	0.000
98	0.0150	0.000	0.000	0.000	0.000	0.000	0.000
99	0.0144	0.001	0.000	0.000	0.000	0.000	0.000
100	0.0138	0.002	0.000	0.000	0.000	0.001	0.000
101	0.0133	0.002	0.000	0.000	0.000	0.001	0.000
102	0.0126	0.002	0.000	0.001	0.000	0.000	0.000
103	0.0121	0.001	0.000	0.002	0.000	0.003	0.000
104	0.0114	0.001	0.000	0.002	0.000	0.001	0.000
105	0.0108	0.001	0.000	0.003	0.000	0.002	0.000
106	0.0103	0.000	0.000	0.000	0.000	0.000	0.000
107	0.0091	0.000	0.000	0.002	0.000	0.003	0.000
108	0.0088	0.000	0.000	0.001	0.000	0.000	0.000
109	0.0081	0.000	0.000	0.002	0.000	0.002	0.000
110	0.0074	0.000	0.000	0.006	0.000	0.004	0.000
111	0.0068	0.000	0.000	0.000	0.000	0.000	0.000
112	0.0060	0.000	0.000	0.000	0.000	0.000	0.000
113	0.0054	0.000	0.000	0.000	0.000	0.000	0.000
114	0.0044	0.000	0.000	0.000	0.000	0.000	0.000
115	0.0041	0.000	0.000	0.001	0.000	0.001	0.000
116	0.0030	0.000	0.000	0.000	0.000	0.000	0.000
117	0.0025	0.000	0.000	0.000	0.000	0.000	0.000
118	0.0021	0.000	0.000	0.000	0.000	0.000	0.000
119	0.0009	0.000	0.000	0.000	0.000	0.000	0.000
120	0.0000	0.000	0.000	0.000	0.000	0.000	0.000

Printed by Books on Demand GmbH, Norderstedt / Germany